SEE, MAKE AND DO

IN THE WOODS

PAMELA HICKMAN
Illustrations by Twila Robar-DeCoste

Formac Publishing Company Limited
Halifax

Canadian Cataloguing in Publication Data

Hickman, Pamela.

In the woods

(See make and do)
Includes index.
ISBN 0-88780-412-8

1. Forest ecology — Juvenile literature.   2. Nature study — Activity programs — Juvenile literature.   I. Robar-DeCoste, Twila L. (Twila Lee), 1956–   II. Title.   III. Series.
QH86.H54 1998   j577.3'7   C98–950079–9

Formac Publishing Company Limited
5502 Atlantic Street
Halifax, Nova Scotia B3H 1G4

Distributed in the United States by:
Orca Book Publishers
P.O. Box 468
Custer, WA USA
98240-0468

Acknowledgements:
Formac Publishing Company Limited acknowledges the support of the Department of Canadian Heritage and the Nova Scotia Department of Education and Culture in the development of writing and publishing in Canada. We acknowledge the support of the Canada Council for the Arts for our publishing program.

Author's Dedication:
For my good friends who make camping at Kejimkujik such fun.

The illustrator would like to thank the following people for their cooperation and assistance: Brenan Coldwell, Callista Coldwell, Katrina Coldwell, Holly Hoang, Krista Packer, Nick Power, Andrew Rowe, Jeremy Savage, Ashley Simms, and Andrew Strum.

Contents

Welcome
to the woods

Any walk in the woods can turn into an adventure of new discovery when you know what to look for and where to look. Read on for a suggested list of things to take on your outing, including a make-your-own bug trapper. Learn how to age a tree stump, read trail markings, identify trees by their bark or leaves, find animal homes and look for clues left by unseen wildlife. Try a hike on your hands and knees, go for a rainbow hike or mark a trail for a friend. There's a lot more to do in the woods than just walk.

Before you go

Heading for the woods? Whether you're going for a short hike, a day-long outing, a ramble at the cottage or camping for the weekend, make sure your woodland adventures are safe for you and for the plants and animals that you are visiting.

SAFETY TIPS

1. Hike in groups of two or more. Children should be with an adult on unfamiliar trails, long hikes, and in wilderness areas. Near the cottage or campsite, children should always tell an adult where they are going and wear a whistle in case of emergency.

2. Check the weather forecast before you go. Weather conditions can change quickly,
especially when you are hiking along the coast or in mountains. Carry rain gear and an extra sweater in your backpack when going on a long hike.

3. It's fun to eat some of the wild plants along the trail, but don't eat anything unless you are absolutely sure of what it is.

4. Remember that the woods may be home to large animals,

such as bears and cougars. Take special safety precautions when you're in these areas. Check the list of field guides at the end of this book for more information.

CONSERVATION TIPS

1. Many families enjoy bringing their dog along. Keep your pet on a leash in the woods to stop it from disturbing and possibly killing wildlife. A leash also keeps your dog safe from unpleasant encounters with animals such as porcupines and skunks.

TAKE ONLY
PICTURES
LEAVE ONLY
FOOTPRINTS

Plants and animals are protected by law in national, provincial and state parks.

2. When hiking in the woods, stay on the trail so you don't trample the plants growing nearby. In some provincial, national and state parks, mountain biking is allowed only on designated trails. Check out the rules before you ride.

3. If you're quiet, you may hear and see a variety of wildlife in the woods. Remember, don't disturb nests or burrows.

4. Handle living creatures with care. After looking at them, return them to where they were found. If you roll over logs and stones, always replace them the way they were.

5. Enjoy the wildflowers where they grow, instead of picking them.

6. Pack a garbage bag with you. Carry your own garbage home, and pick up litter along the way to make the trail a nicer place to visit. You'll also be making the woods a better home for the plants and animals.

Basic bring-alongs

Grab your backpack and put the things listed below in it before you head for the woods. Remember, snacks and a drink will help keep everyone enthusiastic and energized, so go prepared.

- ❏ snacks or lunch and fresh water
- ❏ insect repellant (mosquitoes and blackflies can be a nuisance during spring and summer outings)
- ❏ first aid kit
- ❏ an extra jacket or sweater for long hikes
- ❏ field guides, such as birds, insects, trees, wildflowers (see p.63)

- ❏ binoculars
- ❏ a plastic magnifying glass
- ❏ pad and pencil for making notes or sketches
- ❏ a camera
- ❏ matches
- ❏ garbage bag
- ❏ a hat to keep off sun and insects
- ❏ good walking shoes
- ❏ map for unfamiliar trails and compass for wilderness trips

Bring a bug trapper

Make this simple bug trapper to use on your hike. When you've caught some tiny woodland creatures, take a really close look at them in the bug trapper. When you're done, simply release the unharmed bugs where you found them.

You'll need:
- a small, clear jar with a tight fitting lid, like a baby-food jar
- a hammer
- a large nail, or nail set
- 2 pieces of clear plastic tubing, about 21 cm long, or two drinking straws with bendable "elbows"
- a small piece of very thin material, such as cheesecloth
- tape

1. With the hammer and nail make two holes in the jar lid, about 3 cm apart. The holes should be just big enough to fit your tubing snugly.

2. Flatten the sharp edges around the holes with your hammer. This keeps them from cutting the tubing.

3. Put one piece of tubing through each

hole. On the bottom of one tube, tape a small piece of material. This stops the insects from being sucked up into your mouth.

4. Screw the lid tightly on the jar.

5. Look for a bug that is small enough to fit into your tube without getting stuck. You'll find lots of bugs under rocks and rotting logs.

6. Place the tube with the material taped to it in your mouth. Place the end of the other tube near the insect and suck hard. The bug will be sucked up the tube and into the jar, unharmed. Have a good look at it before you let it go.

Take a hike

If you think all hikes have to be on foot, then you're in for a surprise. Try out one of these mini-hikes and let your fingers do the walking. Just sit down under a large tree and explore the ground. You may find several layers of fallen leaves, including the large, whole leaves on top, and the chewed and cut up bits of leaves below. When leaves fall to the ground they become food for many small creatures, like insects, snails and wood lice. The leaves also rot, eventually turning into soil. Look for small bugs and other creatures in and below the leaves. Feel the top layer of soil. It is usually damp and full of the remains of rotting plants.

Now turn and look at the tree. Explore the tree from the ground up, feeling the bark and looking for signs of life. Many insects, spiders and other small creatures make their homes on and in trees. Look for webs and cocoons on the bark or under loose bark. Are there any tiny holes where insects have invaded the tree? Check for burrows around the tree's roots. Explore as high as you can reach and then look up into the branches for signs of nests or nesting cavities.

TAKE A COLOUR HIKE

The woods are full of an amazing rainbow of colours. Here's a fun way to discover them. Begin with a trip to a paint or hardware store and ask for several different paint colour cards available free on display. At home, cut the cards up into individual colours and place them in a paper bag. Before your hike have everyone stick their hand into the bag and pull out an equal number of colour samples. During your hike, each person must look for something natural to match each of their colours. When you find one of your colours, show it to the others.

A LOOK-FOR LIST

Make a game out of a hike or camping trip by giving everyone a list of things to look for in the woods. When you find something on your list, just check it off or write down what you found.

Things to look for

Make a check mark when you find natural objects that fit the words below. use a different object for each word. look for something...

- ❏ Soft
- ❏ Round
- ❏ Sharp
- ❏ Wet
- ❏ Rough
- ❏ Prickly
- ❏ Smelly
- ❏ Slimy
- ❏ Hard
- ❏ Blue
- ❏ Stretchy
- ❏ Pretty
- ❏ Huge
- ❏ Hot
- ❏ Squishy
- ❏ Noisy
- ❏ Tiny
- ❏ Red
- ❏ Heavy
- ❏ Dry
- ❏ Cold
- ❏ Pointy
- ❏ Oval
- ❏ Hairy

Follow the signs

MARK A TRAIL

Here's your chance to mark a trail for your family or a friend to follow. All you need are several strips of brightly coloured cloth, each one about 2.5 cm wide and 40 cm long.

1. In an unmarked section of the woods, tie cloth strips to branches, shrubs, low plants or rocks at about every 15 paces or so. If you put the markers at

Trail markers

If you're walking or riding on a marked trail, you'll see several kinds of signs. Find out what they mean by checking this mini-guide to trail markers.

Note: These signs are found in national parks across Canada. Provincial, state and municipal parks in your area may use different signs to mark their trails.

Back-packing

Tenting

Facilities for Handicapped

Litter Container

Picknicking

Playground

Shelter

Tobogganing

Viewing

Warden Station

different heights it will make the trail more challenging. Try setting up a winding trail instead of just a straight line.

2. Make your trail in a loop so that the followers end up back where they started. You can make the trail markers easy or difficult to see, based on the age or skill of who is following them.

3. Turn this into a game by timing how long it takes different people to follow your trail.

4. Remember to collect all of your trail markers before leaving the area.

Tree ID

You can learn to identify different kinds of trees by looking at their leaves, bark, twigs or silhouettes. With some practice, you'll begin to recognize some of the most common ones. As you visit the woods in your area or across the country, you'll discover more of the hundreds of tree species. Try this puzzle and match the tree silhouette, leaf and/or cone to its owner from the list of species below. *Answers on page 63.*

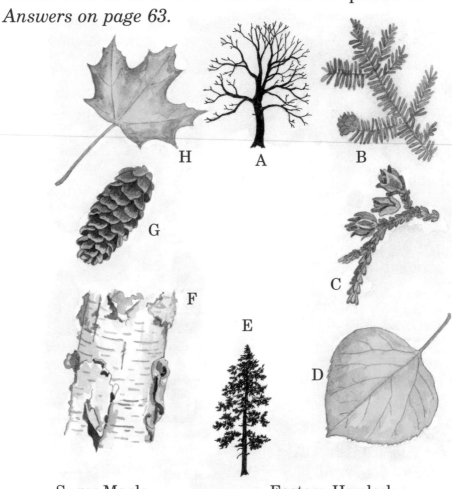

H   A   B

G

C

F

E

D

Sugar Maple
Trembling Aspen
Red Oak
Jack Pine

Eastern Hemlock
White Spruce
Eastern White Cedar
White Birch

Which tree is which?

If you take a walk in the woods in the summer, you'll see some trees with broad green leaves and others with sharp, needle-like leaves.

If you return in the winter, you'll find that the needle-like leaves are still there, but that all of the other trees are bare.

Trees that lose their broad leaves for the winter are called deciduous. Maples are a type of deciduous tree. The needle-like leaves belong to coniferous trees, also known as conifers. Hemlock and spruce are conifers.

In the spring, deciduous trees grow new leaves and produce flowers. When the flowers are pollinated, seeds are formed, often inside fruits. Conifers lose their needles a few at a time, replacing them gradually so the tree is never bare. A conifer's seeds form inside cones.

Trees up close

You may learn to identify a tree by looking at it, but can you tell one tree from another only by touch? Here's a fun game to play with a friend that helps you explore a tree up close.

1. In a wooded area, tie a scarf around your friend's eyes so she cannot see. Lead your friend to a tree and have her feel the tree all over. She should try to remember things like where the branches come off the main trunk, how the bark feels, any irregularities in the tree, the shape of the trunk, etc.

2. Lead your friend about ten paces away from the tree, on a winding route, so that she cannot tell where her tree is located.

3. Take off your friend's blindfold and ask her to find her tree. She may have to feel several different trees before she can locate the original tree.

4. Trade places with your friend and play the game again.

How old is that stump?

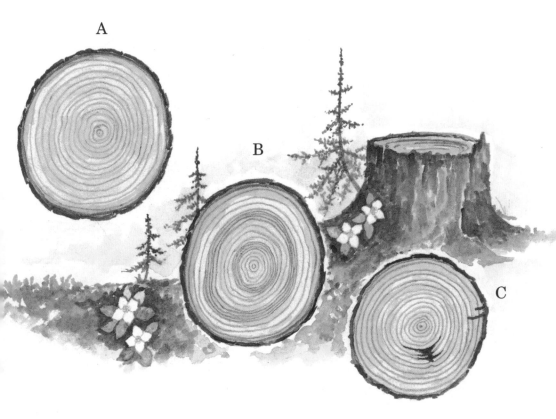

A

B

C

Tree stumps are great for sitting on while you have a snack, but they can also tell you something important about the tree they came from. Look at the pattern of light and dark rings on the surface of a tree stump. The rings are easier to see if you wet the stump first. Each light ring represents how much the tree grew during one springtime of its life. The dark ring next to it shows the tree's growth during the summer and fall. Since a tree grows more during spring, the light rings are the widest. Each pair of light and dark rings equals one year of growth.

Look at the stumps on this page and try to answer these questions. *Turn to page 63 for the answers.*

1. Which stump is the oldest?

2. Which tree grew the most in one year?

3. Find signs of insect damage on one of the stumps.

Anybody home?

Just like people, animals live in different kinds of homes. Some animals burrow underground, others build nests in trees and some make their home right inside of trees, in nesting cavities. When you go for a walk in the woods,

see how many animal homes you can spot. In the spring and early summer many animals are raising new families. If you are careful and quiet, you may be able to watch the adults bring food to their young, or even see the babies poking out of their nests or burrows. Remember, don't disturb the animals or their homes.

Not a nest

If you see a large, ball-like clump of twigs and branches in a black spruce it may be a witch's broom, not a nest. The spruce grows these strange clumps because it has been invaded by a plant called Eastern Dwarf Mistletoe. The mistletoe is robbing nutrients from the spruce tree. Witch's broom may also be found in tamarack and red and white spruce. Other causes of witch's broom include insects and fungi.

Animals out of sight

The rustle of leaves, the trill of a bird, and the chattering of a squirrel all tell you that the woods are full of wild creatures, even if you can't see them. There are many other clues to hidden animals. Look for some of the signs on these pages the next time you're in the woods.

Black Bears scratch at tree trunks possibly to mark their territory or communicate with other bears in the area. Turn to p. 46 for more on bears

Red Squirrels are messy eaters. You'll find the remains of their meals on stumps, fallen logs and large rocks.

Learn to recognize the tracks of some common woodland animals. They are easiest to find in muddy areas or in snow.

Each kind of woodpecker makes a different pattern of holes. These ones belong to a Yellow-bellied Sapsucker. For more about woodpeckers, see p. 40.

If you find a solid, dry clump of fur and bones (about the size of a small sausage) beneath a large tree, you may have discovered an owl pellet. It contains all of the bits of food the owl could not digest, and so spit up.

Look for the droppings of different animals along the trail.

On the ground

When you're out in the woods, don't forget the amazing world of plants and animals at your feet. Take a break from your hike to get close to the ground. Lift up a rotting log, search through layers of leaves, poke into mats of mosses and discover the double life of lichens. Find hidden bugs by setting up a tin-can trap or a bait trail to attract night-flying insects. When you get home, turn some of your finds into art by making leaf prints or a spore print from a mushroom. On your next hike, remember to look down, and watch your step!

Life in the leaves

There is more to leaves than pretty shapes and colours. Leaves are home to a great variety of tiny creatures. When you're in the woods, scoop up a handful of leaves and take a really close look at them. You may be surprised at what lives on and in leaves. Here are a few things to watch for:

• a see-through pathway in a leaf. Leaf miners are tiny insects that burrow between the upper and lower layers of a leaf and then eat their way around, leaving a trail as they go.

• a rolled up leaf full of silken threads. Some moth caterpillars spin their cocoons inside rolled up leaves and stay there until they are ready to emerge as adults.

• ball-like growths on oak leaves. These are called oak apple galls and are caused by a tiny insect that has invaded the leaf. There are many different sizes and shapes of galls found on other kinds of leaves.

• chewed and holey leaves. Many kinds of creatures feed on leaves including caterpillars, beetles, slugs and snails.

• tiny eggs attached to the underside of a leaf. Some insects lay their eggs under leaves to hide them from predators. The young usually feed on the leaves when they hatch.

Leaf art

You can make leaf-shaped borders for your bedroom walls, decorate wrapping paper, create one-of-a-kind greeting cards or paint a whole picture using interesting leaves and some paint.

You'll need:
- leaves
- different colours of acryllic paint
- a paint brush
- paper or blank cards
- water for clean-up

1. Collect different shapes and sizes of leaves.

2. Paint the surface of a leaf.

3. Carefully press the leaf firmly down onto your paper, card or whatever it is you are decorating.

4. Lift the leaf, being careful not to smudge your print.

5. Use several different kinds of leaves to make a collage effect, or choose one leaf and create a repeating border. The sky's the limit.

Chocolate leaves

Here's a yummy use for leaves. Simply melt some semi-sweet chocolate in a pot over very hot water, stirring constantly. Stir in 2-3 drops of peppermint flavouring (optional). Using a paint brush, paint the melted chocolate on the backs of small, non-hairy, non-poisonous leaves such as rose, orange or ivy. Leaves with prominent veins are best. Put the leaves on a tray lined with waxed paper and freeze them. When set, gently peel the leaves off the chocolate, beginning at the stem end. You can use your chocolate leaves to decorate a dessert, or just eat them as is. Enjoy!

Lift up a log

Where can you find salamanders and snakes, centipedes and slime moulds? The answer lies in and under a rotting log. You probably pass by and jump over all kinds of fallen, rotting logs while in the woods.

Next time you see one, stop and take a closer look. Carefully roll over the log and take a peek at what lives underneath. Poke your fingers into the spongy wood and see what's hiding inside. Put your magnifying glass over the thread-like fungi that are spreading in and over the wood. Check out these pages to help you identify some of your discoveries. Remember to put the log back the way you found it when you are finished.

Fallen logs may also be used as "nurseries" where new trees sprout. Yellow birch, for example, have a hard time growing up through the thick layers of leaves on the ground. Seeds that land on a fallen log, however, get a head start and can grow much more successfully. Look to see if any new trees have taken root on your fallen log.

Sow bug

Centipede

Yellow-spotted Salamander

Slime Mould

Step Moss

American Toad

Nature's recyclers

Some animals, such as snakes and toads, use fallen logs for
shelter, as well as places to find food. Others, such as slugs,
wood lice and fungi, feed on the rotting wood. They are helping
to turn dead trees back into soil nutrients that will help other
plants grow. These creatures are known as nature's recyclers.

Marvellous mushrooms

If you go exploring in the woods in the fall, especially a day or two after it has rained, you may find a number of different kinds of mushrooms or fungi growing on the ground, on fallen logs or on live trees.

Mushrooms are plant-like organisms but they have several differences.
- instead of roots, mushrooms have thread-like hyphae (pronounced hi-fee).
- the main part of the mushroom (what you will find in the woods) is called the fruiting body.
- mushrooms don't produce flowers, seeds or fruit. Instead they have tiny, dust-like spores that are spread by the wind. A spore can grow into a new mushroom.

Fly Agaric

- unlike green plants that need sunlight in order to grow, mushrooms can grow in complete darkness. That's because they get all of their food from the surface (soil, rotting log, etc.) on which they are growing.

Take the time to do some "mushroom watching" and you'll be rewarded with a huge variety of shapes, sizes and colours. The names are fantastic too — Carnival Candy Slime, Blue Cheese Polypore and Moose Antlers to name a few. Here are some common fungi to look for.

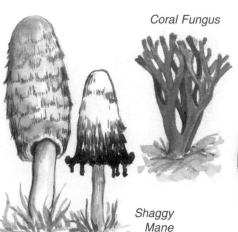

Coral Fungus

Shaggy Mane

Moose Antlers

Chanterelle

Mushroom printing

You can have fun making interesting pictures, called spore prints, with different kinds of mushrooms. The spore prints can also help you identify the mushroom you used.

You'll need:
- gloves
- a mushroom
- a small paint brush
- a piece of black paper
- a plastic container (yoghurt or margarine)
- non-aerosol, fine hairspray
- field guide to mushrooms (optional)

1. We recommend that you wear gloves when handling wild mushrooms. Pick a mushroom and bring it home carefully.

2. Gently break the stem off of the mushroom. Place the cap, gill-side down, on your paper.

3. Cover the mushroom cap with a plastic container and leave it to sit overnight. The container keeps the spores from being blown away.

4. In the morning, slowly remove the container and mushroom cap. You should find a pattern, or spore print. The pattern was created when the spores fell out of the gills directly onto your paper.

5. Spray your spore print with a fine coating of hairspray to keep it from blowing away.

6. Look for a spore print chart in a field guide to mushrooms to help you identify the kind of mushroom you used.

CAUTION

Although mushrooms are fascinating to look at, some of them can be dangerous if eaten. Never eat a wild mushroom unless an expert has positively identified it first. Many different kinds of mushrooms look alike, and a poisonous one can be easily mistaken for a harmless one. If you touch wild mushrooms, always wash your hands thoroughly after you are finished. Here are some mushrooms which you should avoid. Refer to a good field guide for a more complete list *(see p. 63)*.

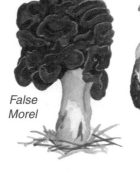

False Morel

Destroying Angel

Mats of mosses

In damp, shady woods the ground may be covered by a thick mat of rich green mosses. Take a close look at some moss and try to find the basic parts illustrated here. Check out the mosses on the ground, on tree bark, rocks and even underwater in shallow streams. You'll soon discover that mosses come in many different shapes. Some look like tiny pine trees while others resemble miniature ferns.

Mosses are sometimes called "pioneer" plants because they are one of the first plants to grow on bare ground or rock. Once they take hold, they create a mat where seeds from other plants can grow. Over many years, a layer of soil builds up that can support much larger plants. Mosses are also important because they protect bare soil from being blown away by wind or washed away by heavy rains. They can soak up water and help keep areas from drying out between rainfalls.

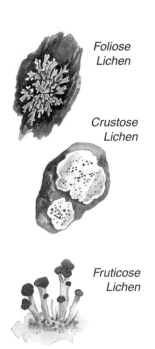

Foliose
Lichen

Crustose
Lichen

Fruticose
Lichen

LOOKING AT LICHENS

Have you ever noticed a greyish-green, crusty growth on tree bark and rocks? What you're looking at is a plant-like organism called lichen. Lichen is actually two organisms combined—a fungus and an alga that live together and help each other survive. With your magnifying glass, take a close look at some lichen. The greenish part is the alga that lives inside the fungus.

Lichen comes in three different forms: leaf-like, or foliose; crust-like, or crustose; and stalked, or fruticose. Look for different kinds of lichen on tree trunks and branches, rocks and fallen logs. Use this mini-guide to identify which kind of lichen you are looking at.

Traditional uses of mosses and lichens

If you lived over a hundred years ago, you would have had many uses for mosses. Pioneers filled the cracks in their log cabins with a mixture of moss and clay. Laplanders stuffed pillows with moss and native North Americans used Sphagnum moss in their babies' pants, like diapers. Some countries still use cut and dried peat (thick layers of partly rotted moss) as fuel. Garden centres sell peat moss as a soil conditioner. Lichens were traditionally used as a source of natural dye and an extract of lichen is still used in food-colouring. Iceland moss, a kind of lichen, is eaten in some alpine and arctic regions.

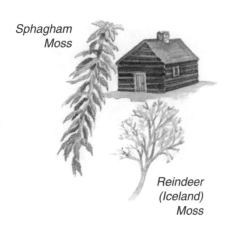

Sphagham
Moss

Reindeer
(Iceland)
Moss

Spore
Capsule

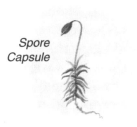

Spring wildflowers

Have you ever wondered why many woodland flowers bloom so early in the spring and then seem to disappear for the rest of the year? You'll find spring-beauty, red trillium, hepatica and bloodroot flowers in May, but by the end of June they are gone. Spring flowers are in a hurry to go through their life cycle for a good reason.

Bunchberry

Spring Beauty

Round-lobed
Hepatica

Plants need sunlight in order to grow. These woodland wildflowers grow under large trees, so they must do all of their growing before the leaves come out on the trees and block the sunlight. As the days get longer and the ground thaws, the plants sprout from underground roots that have survived the winter. In a few short weeks they grow, flower, produce seed and store enough food in their roots to help them grow again the following spring.

A SNEAK PREVIEW

If you have woods near your house or cottage, look for early blooming wildflowers in the spring. Record where you find

Red
Trillium

White Trout Lily

Bloodroot

them by sketching a small map. Alternatively, leave some kind of marker that will help you find the spot again in the winter when the snow might be deep. Return to the spot in the winter with a trowel. Gently dig away the snow and layers of dead leaves until you come to the remains of the wildflowers. Some flowers keep a rosette of ground-hugging green leaves all winter long. The new flower sprouts through the centre of the rosette. Once you've had a look, be sure to cover the plant back up so that it is protected from the cold.

Some woodland animals up close

The woods are home to a wonderful variety of creatures, from the smallest salamander to the biggest bear. Get to know a few woodland animals better. Learn how to find a salamander and bring it home (but only for a little while). Discover the secrets of successful woodpeckers, learn how "flying" squirrels really get around and why deer lose their antlers. And if you're in bear country, take a few precautions to keep both you and the bears safe.

Where are the bugs?

Almost every variety of plant you see in the woods provides food or shelter for a kind of bug. Many other insects live in the soil. When you're out in the woods you may not notice a lot of insects since most of them are well-hidden and some only come out at night. With a bit of home-made bug bait, you can tempt some of these creatures out of hiding. Read on to find out how.

A tin can trap

Here's an easy way to find out what's lurking under the leaves or hiding in the soil.

You'll need:
- a trowel
- a rinsed-out can, such as a soup can, with the lid removed
- some jam or other sweet, sticky substance

1. In some woods, dig a hole in the ground just large enough to fit your can.

2. Place the can in the hole so that the open end of the can is level with the ground surface.

3. Place a small amount of jam in the bottom of the can and smear some more jam around the inside of the rim.

The sweet smell of the jam will attract ground insects and other tiny creatures. They should climb into the can to feed.

4. Leave the can alone for about 20 minutes and then come back to check what has crawled in. After you have had a close look at the visitors, let them go and fill in your hole.

5. Try placing the can in different parts of the woods or under different kinds of plants. Compare the number and kinds of bugs you catch in each place.

A tasty trail

You can attract moths and other night insects by setting up a trail of feeding stations on a summer's evening.

You'll need:
- a bowl
- a fork
- some over-ripe fruit, such as mushy bananas
- some sugar or molasses
- some fruit juice
- an old paintbrush
- a flashlight

1. Mix up some fruit, sugar and juice in your bowl.

2. At dusk, paint a small amount of this mixture on the trunks of trees along a woodland trail. Wait about 30 minutes before returning with your flashlight to see what kinds of bugs are feeding on your sweet treat.

Search for salamanders

When you're lifting up logs and rummaging under rocks in the woods, keep your eye out for a salamander—a four-legged, long-tailed amphibian that looks like a lizard but feels like a frog. Salamanders need to keep their skin moist all the time, so they hide from the sun during the day and come out at night when it is cooler and damper.

Like frogs and toads, salamanders hibernate during the winter. Their body temperature depends on the weather outside, so they only become active again when the weather turns warmer. If you are searching for salamanders in the early spring, you may lift up a log and find one that looks like it is asleep. Chances are its body is still too cold for it to move around. You can gently pick up the salamander and have a good look at it before returning it to its cosy home. Don't forget to roll the log back in place. When you find salamanders in the summer they are usually very fast-moving and are much harder to catch.

A day in the life...

You can make a temporary home for a salamander and watch it closely for a day or so to discover how it lives. You must return the salamander to where it was found. If this is not convenient, don't bring the salamander home.

You'll need:
- a 4-L jar or a small aquarium
- tape
- 2 long pencils or sticks
- sand or gravel
- a plastic container with holes punched in the lid
- a trowel
- a salamander
- a piece of a rotting log
- soil from around the rotting log
- fine screening
- an elastic
- a red light or flashlight and red tissue paper

give the salamander a place to hide. There should be some small creatures in the soil and log to feed the salamander for a day or so.

1. Turn the jar on its side and tape the two pencils to it, as shown. This will keep the jar from rolling. The side of the jar is now the bottom.

2. Put a layer of sand or gravel in the bottom of the jar for drainage.

3. In the woods, look for a salamander beneath a rotting log. Gently pick up the salamander and put it in the plastic container while you get the jar ready.

4. Dig up some soil from around and beneath the log and put it into the jar.

5. Break off a small piece of the log and add it to the jar. This will

6. Put the salamander in the jar and cover the opening with screening held on with an elastic.

7. At home, place your jar in a north-facing window, away from heaters and drafts. Keep the jar out of direct sunlight so it doesn't become too hot.

8. The salamander will be more active in the dark. In order for you to see it better, shine a red light on the jar. Because a salamander can't see red light, it won't be disturbed. If you don't have a red light bulb, try covering the end of your flashlight with red tissue paper. After a day or two, return the salamander to where you found it. Replace the piece of log and soil, too.

Woodpeckers at work

You don't have to see a woodpecker to know that it is living in the woods; just listen for a loud tapping noise, or look for its holes in the trunks of trees. The size, shape, and pattern of the holes are important clues to the kind of woodpecker at work. Check out the mini-guide below.

The Pileated Woodpecker, our largest, makes deep, rectangular holes big enough to fit your hand in.

Downy and Hairy Woodpeckers feed in a spiral pattern as they climb a tree trunk.

Yellow-bellied Sapsuckers drill small holes in a pattern of rows. Unlike most woodpeckers, sapsuckers prefer to feed on live trees.

Make a woodpecker feeder

You can attract woodpeckers to your backyard or cottage in winter by making this simple woodpecker feeder. Downy woodpeckers, the smallest, are the most common woodpeckers to see in your yard.

You'll need:
- a small poplar or birch log, about 40 cm long and 10-12 cm thick
- a drill
- a 2.5 cm drill bit
- a screw eye
- some thick twine or flexible wire
- suet (available at grocery stores, or make your own by melting beef fat and then cooling it)

1. Have an adult drill several holes 3-4 cm deep around your log.

2. Attach the screw eye to one end.

3. Stuff suet into the holes.

4. Hang the log from a branch with twine or wire. Try to place the feeder where you can watch it from a window.

Amazing adaptations

Woodpeckers are well-built for banging holes in trees to catch the insects living inside and for excavating nesting cavities. Here's how they do it:

- large, strong, pointed beak
- strong head and neck muscles
- thick skull bones that act as shock absorbers
- a long, thin, sticky tongue with barbed hooks on the end. The tongue snakes its way through insect tunnels inside the tree and traps the bugs. A wood-pecker's tongue can be up to five times as long as its beak!
- strong toes with sharp, curved claws to hang on to the tree while the birds make their holes
- stiff, strong tail feathers to support the birds as they hang on the tree trunk

Nuts about squirrels

You know you're in squirrel country when you hear loud rustling noises in the dry leaves in the woods, feel acorns and pine cones dropping on your head, or hear the scolding chuck-chuck-chuck of an unseen creature overhead. Red squirrels, grey squirrels, chipmunks and flying squirrels are at home in the woods but they may also visit the trees in your backyard or local park where you can get to know them better.

Watching squirrels travel along tiny branches and jump from tree to tree is a bit like watching a circus performance in the woods. How do they do it without falling? Check out the squirrel to the right for all of its "safety features."

- large eyes are important for judging distances accurately
- bushy tail is used for balance in the air
- a squirrel spreads its legs and holds out its tail to slow its fall when jumping from high up
- sharp claws help it hang on tightly when landing

Flying squirrels have a large flap of skin between their ankles and wrists. When it jumps, the squirrel spreads its limbs out, like a parachute, so it can glide through the air. (It can't really fly.)

Squirrel watching

If you're camping or at the cottage, you'll probably be visited by a squirrel or chipmunk looking for a handout. It is illegal to feed wild animals in any provincial, national or state parks but you can watch as they eat local fruits, nuts, cones and mushrooms, sitting up on their hind legs and holding their food with their front feet. Look for the entrances to their burrows beneath stumps and large rocks and between large tree roots. Look for leafy squirrel nests up in trees and the remains of their meals on stumps, rocks and rotting logs *(see p. 20)*.

Antler alert

A visit to the woods at dawn or dusk, especially near water, is the best time to see some of the larger mammals such as deer, moose or elk. This is when they are most active and often head to water for a drink. Check along trails and in the mud at the shore for signs of tracks or droppings.

If you are very lucky you may find the remains of some antlers in the woods. Each member of the deer family has a unique set of antlers. They grow new antlers every spring and summer and shed them in the fall or winter. Only male deer, moose and elk have antlers but both male and female caribou grow them. Antlers are made of bone.

They are covered with furry skin (called velvet) that contains blood vessels to nourish the antlers as they grow. After the fall mating season is over the antlers' food supply is cut off, which makes them eventually break off. The shed antlers are chewed on by mice, porcupines and other small woodland animals.

White-tailed deer

Moose

Mule deer

Caribou

Elk

45

Bear country

If you're heading for the woods in a wilderness area, first find out if there are bears around. Although you are not likely to encounter even our most common bear — the black bear — it is wise to take a few precautions.

- food should be stored in sealed containers to keep bears from smelling it
- if you're camping, store your food in the trunk of your car, never in your tent
- in backwoods campsites, string your food up high between two trees so that a bear can't reach it from either the tree or the ground. Black bears are good tree climbers
- place all garbage in the containers provided, in the trunk of your car, or pack it out with you when you leave
- don't hike near garbage dumps since these areas attract bears

- if you do come across a bear on a hike, back away immediately and keep your eye on the bear at all times. Don't turn your back on the bear. As soon as you are far enough away, the bear should not feel threatened and will likely leave
- if a bear enters your campsite, loud noises usually help to scare it off

If you are in a national, provincial or state park, report any bear sightings to the park staff. This helps them keep track of where the bears are and helps prevent dangerous situations for visitors and wildlife.

Black bears are good tree climbers.

Black and blue

Although most black bears are black with a bit of white on their chest and a brown muzzle, they also come in different colours. In the west, some black bears are cinnamon or honey-coloured and along the Pacific coast, some have a bluish tinge and others are white.

Woodland fun & food

There's nothing like a camp-out in the woods to introduce you to night-time nature. After a campfire, use your home-made lantern to light your way back to the tent. Too noisy to sleep? Find out what's going on with a mini-guide to the night sounds. In the morning, add some wild edibles to your pancakes. Take time for a sensory game with a friend, and make some simple twig crafts to bring home. Have fun, and stay away from poison ivy!

Night sounds

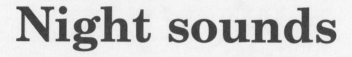

The next time you go camping in the woods, try to stay awake late enough to hear some of the sounds the night creatures make. When you're going to bed, many animals are just waking up and are ready to go out in search of food or a mate. Listen for their different calls, rustlings as they walk or banging as they try to steal food left out by careless campers. Check out these pages to see what kinds of noise-makers you might hear at night.

Bright eyes

When you're in the car at night, have you ever noticed how some animals' eyes seem to glow in the dark when the car's headlights shine on them? Raccoons, deer, skunks, cats and other animals that are active at night have a special, mirror-like layer at the back of their eyes. When light shines into their eyes, it is reflected back out. Since the light passes through their eyes twice, these night creatures can see in the dark much better than people can.

The woods in winter

Exploring the woods is a year-round adventure. Dress warmly and head out to do some winter wildlife watching. In deep snow, snowshoes or cross-country skis make travel through the woods much easier, however, they are not necessary.

WHAT TO LOOK FOR

Tracks

How many different sets of tracks can you find? You can learn to identify the most common ones — squirrels, birds, mice and deer — using a field guide to animal tracks (see p. 63). Follow the tracks to see if they lead to the animal's home. Try to figure out what the animal was doing. Was it being followed, stopping to eat, going quickly or slowly?

Who's hibernating?

A number of woodland creatures keep out of sight during winter. When an animal hibernates, its body temperature and heart rate drop so much that the animal seems barely alive. It stays asleep for several weeks or months and doesn't feed. The woodland jumping mouse is a true hibernator. Many other animals go into deep sleeps during the coldest days of the winter, but wake up from time to time to feed. Chipmunks, raccoons, bears and skunks may become active on mild winter days.

Buds

Since the leaves on deciduous trees have fallen off, this is a good time to take a close look at the buds left behind. Inside the buds are next spring's leaves. Check out the shape, size and position of the buds. Buds and bark are two helpful clues to the identity of a tree or shrub.

Frost cracks and ribs

If you find a long crack or scar in the bark of a tree, you may be looking at a frost crack or rib. When the temperature drops suddenly, especially in spring or fall, the outside of a tree cools down very quickly and shrinks but the inside of the tree stays warm and doesn't shrink. As the bark shrinks, it may split open. In the spring, the tree grows new tissue over the split, forming a scar. Since the new tissue is weaker than the rest of the bark, it tends to split again during cold snaps. Eventually a large scar, or frost rib, forms on the tree trunk.

Nests

Great Horned Owls are one of the earliest nesters, starting in February. Look for their large stick nests high up in big trees.

Colour changes

In snowy areas, weasels and snowshoe hares shed their brown fur in the fall and grow new white fur for the winter. The white colour helps the animals blend in with the snow, or camouflage themselves. Camouflage helps hares hide from their enemies and makes it easier for weasels to sneak up on their prey.

Campfire fun

Whether you are camping, at the cottage or just having a late picnic in a local park, a campfire is a perfect way to end your day. Sing some songs, tell some stories and make a special treat for everyone to enjoy. Check out the tips below and make your next campfire the best ever.

SAFETY TIPS

- build your campfire in the firepit provided. If you are making your own firepit, choose an area away from overhanging branches or tree roots. Surround the fire pit with a ring of rocks to help contain the fire.
- don't break branches off trees or shrubs for kindling.
- small fires are best for cooking since they will burn down more quickly, leaving hot coals perfect for roasting.
- when you are finished, let the fire burn right down. Add water or sand to smother the coals.
- take away all of your garbage.

THINGS TO BRING

- food and drinks
- matches
- kindling and firewood
- a flashlight
- a penknife for sharpening roasting sticks
- a bag for garbage
- a container for carrying water to douse the fire

A Sweet Treat

Roasted marshmallows are a traditional favourite around the campfire. Here's a new idea for marshmallow lovers.

You'll need:
- a large pot and a smaller pot
- a large spoon
- water
- 1 bag semi-sweet chocolate chips
- 1/4 cup milk
- 1/4 cup icing sugar
- marshmallows
- coconut or chopped nuts

1. Boil water in the large pot over the fire.

2. Put chocolate in small pot and place it over the boiling water. Stir chocolate until it is melted. Stir in milk and remove from heat.

3. Add icing sugar and mix until smooth.

4. Dip marshmallows in chocolate mixture and then roll in coconut or nuts.

For variety, try using fresh fruit instead of marshmallows.

A lantern you can make

Before you head for the woods, make this special lantern to use after dark.

You'll need:
- a large tin can with one end removed
- water
- freezer
- marker
- hammer and nail
- candle and matches

1. Draw a design of dots on the can with the marker. The open end is the bottom.

2. Fill the can with water and place it in the freezer until

the ice is solid. Remove the can from the freezer.

3. Using the hammer and nail, punch a hole in each of the dots you made. Make larger holes around the bottom of the can's sides to allow more air in.

4. Outdoors, let the ice melt and pour the water out of the can.

5. Put your lantern over a tea light or pillar-style candle and enjoy the patterns of lights created by your design.

Woodland snacks

If you're out in the woods and have a snack attack but you forgot to pack some food, don't panic—try some wild nibbles. There are many edible treats in the woods if you know what to look for. The most important thing to remember is that you must be absolutely sure of the plant before you eat it.

Some edible plants look very similar to poisonous plants. Don't take any chances. And don't forget to leave lots of snacks for the animals that live in the woods—birds, bears and other creatures also like to feed on berries and other tasty treats.

Here are a few woodland snacks to look for on your next hike.

Blueberries are a favourite treat and can be found in open woods, especially in sandy or rocky soil.

Blackberries grow on long, prickly canes in open woodlands and at the edges of woods.

Strawberries grow in open fields.

Wild about flavour

Put some zip in your sandwich or salad by adding a little wood-sorrel. Throw a handful of wild berries into your pancake batter or add them to your ice cream for a special treat.

Snap off the tender green tip of a yellow birch twig and chew it up to enjoy its fresh wintergreen taste.

Wood-sorrel looks a bit like a shamrock and tastes lemony or sour.

Wintergreen has three glossy green leaves that give off a fresh, minty flavour when you chew them.

The golden-coloured sap leaking out of a spruce tree can be chipped off and chewed like gum.

Twig crafts

Nature crafts are fun to make and nice to have. Here are a couple of projects that you can make with some simple materials. Alder or willow can be found along the edges of wooded streams, ponds or swamps, often growing in large clumps. Remember that it is illegal to cut branches in parks. If you are on private land, be sure to ask permission before you gather any branches.

Make a twig planter

Here's a great outdoor planter for some summer flowers. *You'll need*:

- 6 alder or willow twigs, each about 20 cm long and as thick as a pencil
- 6 alder or willow twigs, each about 12 cm long and as thick as a pencil
- a piece of wood, 1 cm thick, measuring 20 cm x 12 cm
- a hammer
- cigar box nails
- moss
- potting soil
- a few small bedding plants

1. Nail two 20 cm long twigs on top of the piece of wood, so that each twig is resting on the long sides of the wood.

2. Add two 12 cm long twigs, log-cabin style, on top of the longer twigs.

3. Continue to nail the twigs on in sequence until you have used all the twigs.

4. Line the planter with a layer of moss that you collect in the woods.

5. At home, add some potting soil and a few small bedding plants to your planter. Keep the moss moist so your soil will not dry out.

Weave a windcatcher

You can hang your windcatcher from a tree or a balcony and enjoy the colours as it spins in the wind.

You'll need:
- two twigs, each about 25 cm long and as thick as a pencil
- coloured yarn

1. Cross the sticks so that they meet in the middle and tie them together securely with some yarn.

2. Decide on a colour pattern for your windcatcher, starting with the central colour and working your way to the edge.

3. Take a long piece of yarn and tie it near the centre of one of your sticks. Weave the yarn under the neck stick, over the following one, under the next stick and back to where you started. Continue weaving the yarn in this way until you have enough of the first colour.

4. To change colours, cut the yarn and tie the loose end to the next colour and continue weaving.

5. When your windcatcher is as large as you want it, tie the last piece of yarn firmly to one of the sticks and trim the knot.

6. Tie a piece of yarn, or stronger twine, to the top of one of the sticks and hang your windcatcher up where it can spin in the breeze.

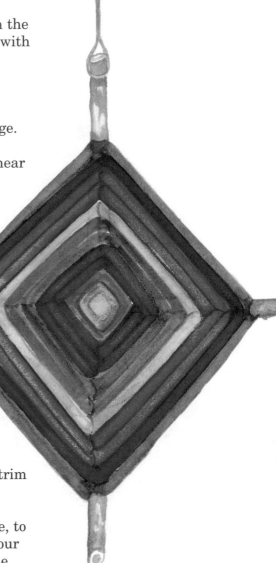

Touch and tell

Test a friend's sense of touch with this nature game. It's made with some simple materials around the house.

You'll need:
- some rinsed-out tin cans, 540 ml or larger, with lids removed
- duct tape
- old large socks
- scissors
- natural objects such as cones, bark, moss, nuts etc.

1. Place some tape around the tops of the cans to cover any sharp edges.

2. Cut the feet off the old socks and discard them. Tape the cut edge of each sock snugly around the outside top of each can. The rest of the sock will form a sleeve, as shown.

3. Find some natural objects in the woods that won't fall apart when handled. For example, pieces of bark and other hard objects work well, but flowers are too delicate. Try to find things on the ground instead of picking them.

4. Place one object into each can without showing anyone. Invite a friend to stick her hand through the sock sleeve and into the can without looking inside the can. Ask her to feel the object and guess what it is.

Leaves three, let it be...

What has three leaves, white berries and grows in the woods? If you said poison ivy, you're right. Being able to recognize this plant could save you a lot of itching and discomfort.

Although it often grows as a low plant in the woods, poison ivy can also be found as a small woody shrub or a vine. The plant's sap contains an oil that can cause small watery blisters on your skin. If you touch poison ivy, wash with soap and water as soon as possible. It's also important to wash your clothes, because the oil can cling to them and then rub off on your skin. You can get a poison ivy rash at any time of year, but it's most common in spring or summer when the plant's leaves and stems are softer and more easily broken, thereby releasing the oil. Even though poison ivy is a nuisance to people, it is an important source of food for many woodland birds, mammals and insects.

Stop scratching

The sap of another plant, jewel-weed, can help relieve the itching of poison ivy. Jewel-weed grows in damp woods and swamps. Just open a stem and rub the juice on your rash.

Suggested Field Guides

Animal Tracks, by O. J. Murie: Houghton Mifflin Co., Boston, 1974.

A Field Guide to the Insects of America North of Mexico, by D. J. Borror and R. E. White: Houghton Mifflin Co., Boston, 1970.

A Field Guide to Wildflowers: Northeastern/Northcentral North America, by R. T. Peterson and M. McKenny: Houghton Mifflin Co., Boston, 1968.

A Guide to Field Identification: Birds of North America, by C. S. Robbins, B. Brunn and H. S. Zim: Golden Press, New York, 1966.

A Guide to Nature in Winter, by Donald Stokes: Little, Brown and Co., Boston, 1976.

Native Trees of Canada, by R. C. Hosie: Canadaian Forestry Service, Ottawa, 1969.

Newcomb's Wildflower Guide, by L. Newcomb: Little, Brown and Co., Boston, 1977.

Peterson's Field Guide to Mammals, by W. H. Burt and R. P. Grossenheider: Houghton Mifflin Co., Boston, 1980.

Summer Nature Notes for Nova Scotians: Woodland Animals, by Merritt Gibson: Lancelot Press, Hantsport, N.S., 1982.

The Audubon Society Field Guide to North American Mushrooms, by G. H. Lincoff: Alfred A. Knopf, New York, 1981.

Trees of North America, by C. Frank Brockman: Golden Press, New York, 1968.

Tree ID
Answer to puzzle on page 14:

A: Red Oak
B: Eastern Hemlock
C: Eastern White Cedar
D: Trembling Aspen
E: Jack Pine
F: White Birch
G: White Spruce
H: Sugar Maple

How old is that stump?
Answer to puzzle on page 17:
1: B
2. B
3. C

Index